WILD and unsubstantiated RANTINGS

Philly Salfield

ALSO BY PHILLY SALFIELD

God, Humanity and the the Mystery of the the Universe – 2008 – available from lulu.com

Where We Went Wrong – 2012 – available from lulu.com

Inside the Mind of an Alien – 2013 – available from lulu.com

I am a Vegetarian – 2015 – available from lulu.com

What Are You? - 2016 – available from lulu.com

Byron Community Primary School – 2019 – available from lulu.com

I dedicate this book to my grandchildren – Ivy and Maverick (Mickey Mouse) and the two on the way – and hope that the pessimism underlying much of this book never comes to pass.

Written by Philly Salfield
Cover design by Philly Salfield
Published by Philly Salfield
Philip Salfield asserts his moral and legal rights to all the wordings and images contained within this book.

Published 2020

CONTENTS

	PAGE
The Octopus	11
Beam Me Up Scotty	13
Gaia – Climate Change – COVID-19	16
The Biggest Mystery in the Universe	18
Gaia – Over-population – COVID-19	21
Dr Cohen	22
Corona the Virus Not the Beer	26
Not Coming To Terms With Death	30
A Numerical Perspective On Our Wonderful World	34
Why Are Humans at the Top of the Evolutionary Ladder	36
Black Hole	38

Deus Ex Machina	41
Finland Leads the Way in Education	42
Jack and his Packet of Crisps	44
Ghost TV	45
Pithy Proverbs	46
Kuru	48
The Persistence of Family Traditions	49
Depression With No Discernible Cause	52
On Being Mildly Jewish	53
Out Of Sync	57
BOMDAS and PEMDAS	59
Silly Salfield Interchanges	61
A Visit From Stella	65
COVID Appraised Statistically	66

The Standard Model of Physics	68
Things I Don't Believe In	69
Unsolved Mysteries of the Universe	70
White Couple Making Love	71
The Case From Hell	72
Death in a Time of Corona	77
Do I Believe in God?	79
My Final Comments on This Fucking Virus	84
The Brain's Way of Healing	85
Reductionism Wholism and Mu	88
Consciousness Beyond Life	91
Do We Really Exist?	92
The Universe in Your Hand	94
Dorothy's Funeral	97

Biocentrism	**99**
Major Historical Events of the Modern Era	**110**
A Picture Tells A Thousand Words	**111**
The Author	**114**

THE OCTOPUS

Good Friday is the day on which the crucifixion of Christ is commemorated in the Christian Church. It is traditionally a day of fasting and penance. This is Good Friday at Sydney fish markets ..

In the wild, the octopus is actively discovering his environment, not waiting for it to hit him. The animal makes the decision to go out and get information, figures out how to get the information, gathers it, uses it, stores it. This has a great deal to do with consciousness. Philosopher Peter Godfrey-Smith has given this a great deal of thought, especially when he meets octopuses and their relatives, giant cuttlefish, on dives in his native Australia. "They come forward and look at you. They reach out to touch you with their arms." They have brilliant minds with less of the centralised self of the human mind.

BEAM ME UP SCOTTY

The next huge development in affluent human society will be tele-transportation. It will not replace the internet or phones, it will replace planes, trains, buses and to a large extent cars. The implications for human society are enormous. Mostly things will be better. Countries will cease to exist but unfortunately warfare will not as there will be massive gaps between the affluent with tele-booths and the poor without and so there will be terrorist and class-based warfare on a mass scale but no wars between countries.

This is how I envisage it. Anyone now who is affluent enough to own a car will have their own tele-booth at home – it will probably be as expensive as a car but for most people it will replace having a car with virtually no running costs. The first thing it will replace will be planes. You will enter your tele-booth at home and at the same speed as a phone call or the internet you can travel to anywhere in the world that has a tele-booth – shopping centres, hotels, friends' houses, beach resorts, national parks, office blocks, factories, any large work place etc. How do you get to work? Anyone with a tele-booth at home and a tele-booth at work will get there virtually instantaneously, so no more commuting, no more commuter traffic. But where will you live and work …. anywhere in the world ... you might live in Sydney and work in London or live in Byron Bay and work in New York etc etc. ....And where will you have your lunch if you work in London, well you could have lunch in Tokyo or Byron Bay or anywhere with a tele-booth. So where would you live .... well, you could live in the Himalayas or the Bahamas or anywhere you like, you would still get to work instantaneously.

So now we begin to see the enormous implications – what will happen to where people live and work – what will happen to cities, what will happen to beautiful currently sparsely populated rural areas and what will happen to countries. As I have said countries will disappear because you can live anywhere in the world and work anywhere in the world and travel anywhere in the world so boundaries will be meaningless. Bikes will become much more popular – you will pop into your home tele-booth with your push-bike, go anywhere in the world with a tele-booth and then go for a bike ride or cycle to a walking track and go for a walk or cycle to the beach for a swim. How will you do your shopping? You will go straight from your home tele-booth to the shopping centre, do your shopping, get back in the shopping centre tele-booth with your shopping and take it straight home.

Most people in the world do not go for country drives or country walks or for that matter ever go outside urban areas so for most people the tele-booth will replace their car completely. Other people will take to tele-porting around with their push-bike to explore local areas or get to more obscure walking tracks. The rich will of course still have a car of some sort as well.

As for commercial transportation there will be huge tele-transporters to deliver to retail premises and trucks will be used only for impoverished outlets without tele-transporters. So, roads both between and within cities will virtually disappear – there will just be single lane roads and pedestrianised cities. However, small outlets in the affluent world will nearly all disappear – people will not go to shops outside shopping centres when they can tele-transport directly to the shopping centre, even if it is only for a carton of milk.

Tele-transportation will create a greater change to human societies than there has ever been with any other innovation. Life will be better in many ways. It will certainly be a huge improvement over the use of slow polluting transport. The trend towards virtual reality will diminish as people will walk and cycle in the real world but how it will affect cities is hard to visualise.

Gaia – Climate Change – Covid-19

Gaia is our home. It comprises a molten inner core, polar ice-caps and an atmosphere surrounding the earth. Its surface is mainly ocean with a superficial veneer of land. It is on this land that we, Homo sapiens sapiens reside, along with many other wonderful species of plants and animals. We have not been good news for Gaia nor for the other species of earth. We have had a significant negative impact and we have the completely mis-guided hubris to think that we are the masters. If the temperature rises by the maximum prediction of 6°C, it will be a mere blip, a mere blink, perhaps the best analogy is, a mere teardrop in the life of Gaia. Gaia has experienced much higher and much lower temperatures. If Homo sapiens sapiens self-destructs, the same applies, a mere blink in Gaia's history. Gaia is an integrated system of such complexity within itself and with its relation to the rest of the universe that climatologists and politicians are utterly naïve if they think that they can predict what effect the emission of carbon will have on the future climate. They are even more naïve if they think that they can bio-engineer it back to some state in the past. There is no going back. There is only moving forward. And they are finally utterly and completely naïve if they think that Gaia will tolerate seven or more billion of the species Homo sapiens sapiens for long. Gaia will always move towards balance until eventually Gaia will die and revert to the entropy of the universe, but this will happen through the power of universal energy and not through any doings of one unfortunate little species.

I wrote the above well before the COVID-19 pandemic. You would think that everyone would now hear Gaia screaming, but it would appear not. Homo sapiens sapiens thinks in terms of decades whereas Gaia operates in terms of tens of millions of years. It's not a question of who will win, Gaia or Mankind, Mankind is one small part of Gaia, a part, not an opponent and definitely not a master. This pandemic was one little foray into reducing our numbers and mankind has responded in the worst and most unenlightened manner possible. This will make the next series of crises even more devastating until Mankind realises that we have to reduce our numbers by a vast amount in order to reduce our terrible negative impact on our home. We can do it the hard way or the devastatingly hard way and the monumentally intelligent species that we are has decided to do it the devastatingly hard way.

THE BIGGEST MYSTERY IN THE UNIVERSE

There are many mysteries in the universe but the biggest ones are : the origin of the universe : the origin of life : the mechanism of evolution. There are many others such as the nature of consciousness, black holes, dark energy and superpositionality in the quantum world. However, one tends to forget that the factor that many of these mysteries have in common and the most puzzling phenomenon of all is "time". If we could understand time then many of these other mysteries would become answerable. Unfortunately, we will never solve this mystery for the following reasons. Without time, nothing can begin or end and yet the first and biggest mystery is that we believe that the universe did begin even though time did not and that the universe can never end even though everything must have an end. This is the ultimate oxymoron. We humans constructed the artificial phenomenon of time; for something to exist it must be open to perception by some sense or other and yet there is no sense which can perceive time. This can then be extrapolated to absolutely everything; every thing must have begun and everything must end and yet, thanks to the ultimate oxymoron of the universe having no beginning and no end, it is impossible for anything to have a beginning or an end.

The universe never began, I say this despite the fact that most physicists today with their brilliant minds, agree that it did begin. It is my contention that it never began because if it began, what was there previously .. nothing and no time! .. but also it must have begun because if it didn't begin how did it get here! The only logical conclusion to these two impossible premisses is that there is something faulty in the whole notion of time and hence in the notion of beginnings and endings.

We humans can and do understand the mysteries of the universe but not in our conscious waking state but whilst asleep or meditating or unconscious; in these other states of consciousness, there is no mystery. But there is no way to translate this understanding in our other states of consciousness into this current conscious state of wakefulness in which I am writing and in which you are reading.

However, from the foregoing argument, we can logically conclude that the universe had no origin in time, hence nor did the earth, hence nor did life. There is an alternative explanation in the theory called biocentrism, whereby our consciousness brought into play everything else in the universe. But we cannot comprehend this, the term for this being "the ineffable".

I have nearly finished this wild rant except to say that although we cannot comprehend doing without time, which as I say is the main erroneous and obfuscating factor to us understanding the mysteries of the universe, the concept that time is an artificial construct is bolstered when we consider the measurement of time. Time can only be measured if one thing is compared in relation to another e.g. daylight and seasons on the earth due to the relationship of the earth to the moon and the sun. But there is no fixed static point anywhere in the universe by which to make comparisons .. not the sun, not the earth, not the moon, not the lifespan of radioactive atoms, everything is constantly changing, so how can time possibly be measured in any meaningful, consistent manner .. and this should not surprise us now that we have established that time does not actually exist in the first place!

GAIA – OVER-POPULATION – COVID-19

There are many critical issues facing the world today : climate change, environmental degradation, species extinctions, racism, warfare. The most crucial of all is the massively over-represented population of the human species, 7.5 billion trending towards 9 billion. This number desperately needs to be reduced not by half, not by a quarter but down to maybe 500 million i.e. for every 15 people in the world we should be aiming for one person. A population of this size would save the environment, eradicate racism and finish with large scale warfare. Gaia attempted to start this process with COVID-19 and instead of flowing with it we humans have fought it tooth and nail and probably will beat it, thus setting the scene for more and more catastrophes. Even before COVID is over there have been race riots in America whilst China and America are flexing their weapons at each other. There will be more pandemics and Gaia will continue to produce catastrophes until a reasonable population of human beings is achieved. If only we the species had enough intelligence to go with Gaia and to work towards this goal it could all be achieved with so much less suffering but we haven't and so it won't.

DR COHEN

Dr Cohen lives in a large luxurious apartment in the Kurfürstendamm in the heart of Berlin. His consulting rooms are situated below at street level. He has a full register of female patients many of whom adore him and all of whom respect him wholeheartedly. He would never be seen outside without a jacket, tie and hat. He is the epitome of German respectability.

From late 1938 he started to lose all of his non-Jewish patients; they averted their eyes when they encountered him and by the outbreak of war in 1939 his remaining patients were all Jewish and they could only pay a token fee or nothing at all. He somehow survived on this and by selling his valuables and by surreptitious handouts of food from some of his old acquaintances.

This continued until the day in 1941 when the Gestapo dragged him out of his apartment, knocked him to the ground and kicked him repeatedly in the head and body. He lay there unable to move and could do nothing as they dragged his wife out and ripped her dress off, accompanied by bawdy and derogatory comments and laughter; they bashed and kicked her too. The next thing he saw was his two year-old daughter being held by the ankles whilst the Gestapo officer banged her head repeatedly against the concrete pavement until she lay in an unrecognisable motionless state. Then his four year-old son appeared, screaming as he was carried over the shoulder of another black-uniformed officer before he and his mother were taken away by two of the Gestapo; Dr Cohen never saw his wife or child again. The body of his two year-old was left where it lay whilst Dr Cohen was thrown into a cart and delivered to the railway station, other Jews being collected along the route.

Dr Cohen found himself in the carriage of a goods train packed to the hilt with other Jews. They embarked on a three day journey in freezing conditions. A few bites of bread and a few sips of water were his only nourishment. The carriage was completely bare. There were puddles of piss and piles of shit. When he slept, crushed alongside others, he was surrounded by excrement.

Upon arrival at the camp he was subjected to the worst thing which he had ever experienced; worse even than the sight of his wife being brutalised and his adored baby daughter being beaten to death; it wasn't the freezing cold nor more blows nor the hunger; it was being forced to strip naked in full view of all the other prisoners: men, women and children; the Gestapo officers screaming in his face and battering him until he complied. It was then that Dr Cohen ceased to exist. His ego, his identity, his emotions had all been tied up with his profession, his status, his material well-being and of course, his wife and children. From then on he became an automaton without a soul, programmed to somehow survive.

After he had stripped off his smart consultant's clothing, he quickly grabbed some of the items of ragged clothing which the guards were throwing at them and which he was now to wear for the duration of his life in the concentration camp.

Dr Cohen's profession was recognised by the Gestapo and his medical skills were utilised. As group after group of the Jews were taken for their "showers" in the gas chambers, his job was to search through the corpses and check that none were still living, which miraculously happened occasionally. His job was then to administer a lethal injection to the survivor. He then went through the corpses and extracted the gold from their teeth.

He was given small amounts of extra food in recognition of his services, used his intelligence and medical knowledge to successfully avoid contracting any of the rampant terminal diseases such as scarlet fever and diphtheria and, even though he no longer existed as a human being, his instinctive survival system somehow enabled him to be one of a tiny minority to survive the next four years in the camp.

When the Nazis were defeated in 1945, the camp guards and administrators all fled and the remaining prisoners were now free. They were rescued by the American liberation army and taken to temporary hospital facilities. Within a few days of nourishment and rest, Dr Cohen was able to function normally and the Americans appointed him to carry out medical duties for the survivors.

He donned the familiar white coat and went to check on the supplies available in the pharmacy store-room. He unlocked the door, entered and locked it from the inside, opened a bottle of barbiturates, swallowed the lot, swilled them down with water and lay down on the store-room floor.

CORONA THE VIRUS NOT THE BEER

Fast forward to 2021. People in formerly democratic countries come to realise that they are now living in an authoritarian state. We are informed that in order to safeguard from future viruses there will be compulsory viewing of daily announcements of the regulations on how we are to live our lives. Mass protests or demonstrations are now illegal. Dissenters will be arrested.

The people of the allied nations in my parents' generation gave up their lives and suffered incredible hardships in order to defeat the tyranny of Naziism. They fought for democracy, for the rights of individuals to pursue their lives as they thought fit, to allow for individual expression of opinions, to be allowed to demonstrate against their government and to be subject to only the fair and just implementation of the law.

Through the mass fear of an influenza virus and orchestrated by the world's leaders and the mass media, we have now let ourselves be subjected to tyrannical rule. We are not allowed to leave our houses or convene in groups or be too close to other human beings. We are not allowed to travel internationally and we now have border patrols and controls within Australia!! Where did these draconian powers suddenly come from within our democratic countries which uphold the values of individual freedom. Who knew or even guessed that they had such powers. Within only a few weeks the majority of the population meekly acceded to living in an authoritarian state. Yes, we still have freedom of expression, but already people are baying for the blood of anyone who transgresses the rules. People are vilified for expressing any views which are not politically correct concerning the corona virus. Who knows but before long subversive comments will be made illegal and perpetrators will be arrested.

The argument is that all measures must be taken to stop the spread of the virus no matter what incursions on civil liberties. I don't believe that the prevention of the harm from the virus outweighs the loss we have seen to the fundamentals of our society. Furthermore, I believe that there will be untold amounts of misery and death resulting from the draconian measures in place which will outweigh the ones incurred by the virus itself. What is more, the measures do not even purport to eliminate the virus but merely to slow its spread. Is that a good thing? It will mean that the pandemic continues for a much longer time .. because it is slower .. it will mean that the virus does not actually burn itself out but will linger on and recur in full force. Whereas, if it was left to burn itself out, then it would never recur. So the draconian measures which have played havoc with all of our lives have indeed slowed down the spread of the virus. This has enabled medical services to better cope and alleviate suffering. But this is an apocalyptic situation, there is no way of preventing death and suffering, there is only balancing up which approach will cause the least and it is my opinion that the present approach will have caused the most.

By good fortune, the grounding of the airlines, the quarantining of people in their homes and the lack of consumer spending, will all be good for the environment, in fact probably be more effective than the small measures which governments had been making. But the point is that the environment is infinitely more important to the survival of the human race than the danger from any virus. The vast sums of money that have been put into slowing down the corona virus would have provided far more benefit to humanity by being put into environmental improvements. So, why has so much more money been spent on anti-virus measures than on environmental measures? - Because the corona virus is a nice simple issue whereas the environment is a very complex one and too nebulous and long-term for most people to get their heads round and therefore not of so much interest to the power-hungry politicians in charge.

Interview with Prof Dolores Cahill on youtube

The first ten minutes is her citing her considerable and impressive qualifications as an immunologist. She explains why the process of lockdown and social distancing is not only unhelpful but that it is also counter-productive. The virus poses no big issue for anyone under 50 and in good health with no pre-existing conditions. It is important for your immune system to come in contact with microbes and viruses. The sensible procedure would have been to quarantine vulnerable people only not healthy people. They should have then been treated with Vitamin D, Vitamin C, zinc, hydroxychloroquine and good nutrition. This will clear the virus and from then on they are immune. One percent only of these people would die and half of those would be over 80. Huge numbers of people were already immune and many more now are. The practice of social distancing is counter-productive to herd immunity.

NOT COMING TO TERMS WITH DEATH

Now that I am 70 I have become obsessed with thoughts of death. My own demise primarily but also of all my other septuagenarian friends and family. I think we were all brought up on the notion of our allotted lifespan being three score years and ten and although that no longer applies the fact still remains that I do not have many years left. People have said to me things like 70 is only a number, you are only as old as you feel. But let's be honest, this is a load of shit - once you are 70, no matter how good your health is or how good your family genes are, the fact remains that you do not have many years remaining.

Here are the ages at which some great people died :

Einstein 1879-1955   76
Vincent van Gogh 1853-1890   37
Leonardo da Vinci 1452-1518 66
Shakespeare 1564-1616   52
Beethoven 1770-1827   57
J S Bach 1685-1750   65
Mahatma Gandhi 1869-1948   79
Marie Curie 1867-1934   67

Plato 428-347 BCE   81

Let's say, for argument's sake, that I have another 20 years, we all know that your first 20 years of life go on for a long time and are extremely significant but the last 20 years go by in a flash and are a complete waste of time. Plus I am likely to be a gibbering decrepit old fool for the last 10 years, with my only interest being visits to the doctor to tamper with an increasing number of ailments.

So, although I may already sound like a gibbering idiot, to the best of my knowledge I am in fact still compos mentis and so, while I am still able, I think this is a good moment to reflect on what I have learned about reality .. not the meaning of life because my friend Paul has already worked out that the only possible meaning is to enjoy it.

But this is what drives me mad with frustration – I and all of us are going to die without ever having any possible explanation as to the nature of reality. Here we are living our lives, taking them seriously, heading towards death and yet none of it makes any sense. And that is of course why many people turn to religion, but this is of no help to those of us who are too clever to think that any of the religions has any helpful explanation as to the mysteries of the universe.

The greatest mystery is how people can be so stupid as to think that the invocation of some god or other somehow solves the absurdity and inexplicability of the whole business. It is also disappointing that most scientists, who I would have hoped had more intelligence, think that the Big Bang Theory is any better as an explanation than a god.

The only two theories I have ever come across which contain some semblance of plausibility are biocentrism and pantheism. However, they only offer the merest glimmer as they do not come to terms with the notions of time and space and there is no way of escaping our notions of time and space and there is no way that time and space can make any sense.

Then we have the questions of my own consciousness and my own soul or spirit. According to the theory of biocentrism, which I find the most plausible, I have created the whole universe as it is out of an infinite number of possibilities, yes and that includes all of history, all of you, everything and so presumably when I die then everything including all of you, stops existing. The stuff about souls and spirits doesn't make any sense unless they exist outside the realms of time and space, which is probably the case but nothing any of we humans will ever be able to grasp.

So there we have it, imminent demise and nothing understood or with any possibility of being understood.

Am I scared of death? - I'm not sure. Am I scared of dying? - I'm not sure. Am I scared of living with pain or disability? - Absolutely. Would I want to live forever? - Absolutely not. Would I want to be cryogenically preserved and re-animated in the future? - Absolutely not!

I should just get on with it, try to take every day as it comes, enjoy it and fuck the absurdity. If only I could!

A NUMERICAL PERSPECTIVE ON OUR WONDERFUL WORLD

There are one hundred trillion neural interconnections in one human brain. Not many numbers can surpass this yet even this is dwarfed by the number of stars in the observable universe. What is a trillion? A billion used to be a million million until the Americans demoted it to a thousand million. Flowing on from this, every other big number has been demoted a thousand-fold so that as you work your way up from a billion to a trillion and so on, they are all a thousand times the previous one and a thousand times smaller than they used to be. But does it make any difference which definition we are using when the number of stars is supposedly seven sextillion .. it makes no difference to me whether there are seven sextillion stars or seven quadrillion. Actually, once you get past a few million it is all the same .. they are all unimaginably huge numbers. So it makes no difference that the definition of all these big numbers changed by a thousandfold. Once anything is that big it may as well be a zillion or a gazillion .. which are imaginary numbers meaning "one helluva lot"! One hundred thousand is a comprehensible number as this is the crowd at a sports event. Up to forty million people attend the Kumbh Mela Festival in India .. any more than that and it might just as well be a quadrillion as a billion or even a gazillion even though the magnitude of difference between some of the numbers is greater than the total number I can envisage in the first place!

The magnitude of a number beyond a few million is irrelevant as it is beyond the imaginative capability of your brain, even though it has one hundred trillion interconnections. Perhaps an ordinal list of sizes is more meaningful. For instance, the universe is fourteen billion years old and our earth is a paltry four and a half billion years old compared to three hundred billion stars in our home galaxy, The Milky Way. Unfortunately, not! If we cite the age of the universe in seconds instead of years, then you would see that its number becomes more impressive than the number of stars in The Milky Way. The internet connects the burgeoning human population of seven billion by two billion devices but these numbers pale into insignificance compared to the ten quadrillion ants and this again is much less than the seven quintillion grains of sand on earth. And none of these numbers come close to those seven billion trillion stars.

WHY ARE HUMANS AT THE TOP OF THE EVOLUTIONARY LADDER

CLEVEREST
octopus
black cockatoo
raven
bonobo
dolphin
   . . .
human

FASTEST
falcon
cheetah
antelope
hare
polar bear
horse
   . . .
human

FASTEST REFLEXES
trap-jaw ant
fly
humming bird
cat
  . . .
human

STRONGEST GRIP
dung beetle
ant
gorilla
eagle
tiger
  . . .
human

MOST POWERFUL
blue whale
white pointer
grizzly bear
ox
elephant
  . . .
human

BLACK HOLE

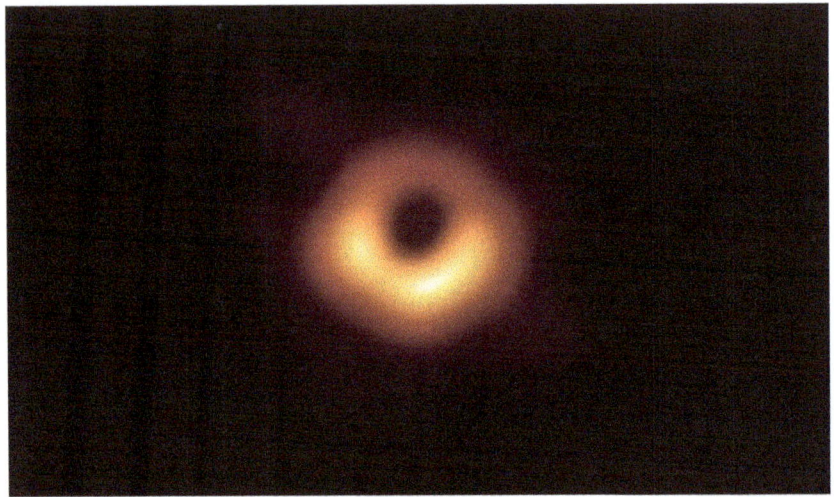

This image of a black hole has created great excitement and is a major news item.

It is supposedly the image of a black hole obtained by concentrating several radio telescopes on a star several billion light years away and using millions of gigabytes of data to produce the image.

If I understand this correctly this means that a couple of billion years after the Big Bang, when stars had been formed, they then started to fizzle out. So several billion years ago, this star reached the end of its life and imploded. This makes sense as atoms are all attracted to each other and if the inside of the star is denser than the outside then once they have lost the energy to shine outwards as a star then they are all attracted inwards. The energy that this involves compresses the matter many billions of times and it is so powerful that it pulls in everything in the vicinity. This almost makes comprehensible sense so far. But one of the things that gets pulled in is light itself. However, we are told that light photons travel at the maximum speed that is possible, then how is it possible that they get pulled in? Furthermore, if the light is pulled in, yes it would make a black hole. But a black hole is the absence of all light so what does it mean to say that you can have an image of it?

It is my opinion that the astrophysicists are not being quite honest and that, despite all their gigabytes and data, that this is still all incomprehensible theoretical conjecture. Black holes along with the Big Bang are considered to be singularities, meaning that they are outside the laws of physics. But a law by definition has to apply to everything otherwise it ceases to be a law. Therefore the conclusion should be : scrap the laws of physics and start again. In other words, we know nothing.

This image has created great excitement not only in the astrophysical world but also in the general public via the media. A huge amount of data and a massive amount of technology and brainpower was required to produce an image which is very similar to what I would have drawn by intuition, does not look particularly exciting and I fail to see what it has added to our understanding of the universe.

DEUS EX MACHINA

Deus ex machina –.. is latin but was originally Greek .. for *god in the machine*. In Greek and Roman dramas they used a machine to lower a god on stage. It means *some unexpected twist to provide a solution to a problem, an unexpected power or event saving a seemingly hopeless situation*. For example, it has been said that Lamarck's explanation of transgenerational memory is deus ex machina, implying that he just made it up to fit his theory .. however, his explanations are now the basis for epigenetic theory. A literary example is in *Lord of the Flies* when the fate of Ralph being killed by the other boys seems inevitable but in a **deus ex machina**, a ship's captain suddenly appears on the beach and he is saved. In Australia there is a motorcycle dealer and clothing company called *Deus ex Machina*. *Diabolus ex machina .. demon in the machine* .. is the evil counterpart .. and also the title of the 2019 book by the brilliant Paul Davies.

FINLAND LEADS THE WAY IN EDUCATION

In my book **Where We Went Wrong**, I advocated an education system based entirely on project work with students choosing their own subjects, no curriculum or subject boundaries and no competition. As far as I am aware the Finnish educational authority has not read my book, but they have come up with an educational system which completely follows my recommendations! Numeracy and literacy are not taught, yet, despite this, ironically, Finland is the leader of all the westernised countries in measures of numeracy and literacy.

The educational paradigm is phenomenon-based learning. This means individualised and independent project work. Holistic real-world phenomena provide the motivating starting point for learning instead of traditional school subjects. The phenomena are studied as holistic entities, in their real context, with no subject boundaries. No subjects are taught separately and there is no curriculum. The aim is to develop critical thinking, creativity, innovation, team-work and communication. Literacy, numeracy and factual learning are incidental.

They start school at seven years old. School hours are 9.30-2.30 with plenty of breaks. According to the OECD, they have the least amount of homework in the world. Finnish students are excelling without having to worry about grades and long hours.

JACK AND HIS PACKET OF CRISPS

Jack calls in at the station shop to buy newspapers and a packet of salt and vinegar crisps. He gets on the train and sits at a table opposite a stranger. Jack opens the bag of crisps which are on the table next to the other man's bottle of water. He pops one in his mouth and notices the other man is just staring unblinkingly at him. Then the other man takes a crisp and pops it in his mouth. Jack maintains his stare at the other man and takes more crisps, and the other man then takes more crisps, and they continue eating and staring unflinchingly at each other until the whole packet is finished. The train arrives at the next station, the other man takes the empty bag, throws it in a bin, reaches up to the luggage rack, grabs our man's suitcase, puts it on the floor and gets off the train. Jack gets up, quite flustered and shaken, picks up his coat …. and finds his bag of crisps in the pocket!

GHOST TV

I bought a chromecast for my tv which enabled me to watch the internet from my computer on the tv using a wireless signal. Danny set it up for me and got it working. A few days later I tried it out. It just wouldn't work. The tv showed an error message of "no signal" no matter which channel I tuned in to or how many times I switched it on or off or tried different things. I had been trying to watch our home video "The Ausalfields" from youtube. I finally checked the back of the tv and the chromecast wasn't there, which meant that Jono must have stolen it. I knew that Danny had one of his own so it couldn't be him. So I went round to Jono's flat (next door) and found Jono and Sara were completely freaked out. The tv programme they had been watching had suddenly been replaced by "The Ausalfields". They were speculating about poltergeists and all sorts of weird supernatural phenomena. The explanation was that Danny had stolen my chromecast and put it into Jono's tv so they could watch youtube at the party a few nights earlier. Neither Jono, Sara nor I knew anything about it. Jono and Sara didn't even know there was such a thing as chromecast. Unbeknownst to any of us, I had been controlling what they were watching on their tv from the computer in my flat!

PITHY PROVERBS

Proverbs or aphorisms are pithy sayings which people often quote at you to prove a point. They are usually in the form of an analogy or metaphor. This serves the purpose of obscuring what is meant and making it seem to be profound and significant. In fact, it is making it open to different interpretations. Here are some examples of the nonsense inherent in these proverbs.

Fair words butter no parsnips. **This means that if you are nice you will achieve nothing.**

A bird in the hand is worth two in the bush. **This means that you should stick with what you have or know and not risk doing anything to improve.**

Many hands make light work. **This means that you should get others to help you do your chores.**

God grant me the serenity to accept the things I cannot change, courage to change the things I can and wisdom to know the difference. **The God bit is meaningless, you either have serenity or you don't. You have no choice but to accept things you cannot change so there's no point in asking for**

God's help anyway. If it requires courage to change something you probably don't want to change them anyway and what are these things anyway. I believe not only in Gaia but in universal intelligence and so there is nothing which is impervious to me trying to change it so that shows lack of wisdom. The aphorism finishes off with the trick of language which invokes the idea of wisdom so you are stupid if you don't agree with it.

Live every day as if it were your last. You'll probably end up dead after a few days if you follow this advice.

Live in the moment. You'll probably end up dead even faster.

KURU

Kuru is a neurological disease found only amongst the Fore tribe in PNG. It is either the same as or very similar to Creutzfeld-Jacob disease (CJD) and also bovine spongiform encephalopathy (BSE) (mad cow disease). Australian and American researchers discovered that it was transmitted through eating the human brain of someone who had it but that it was not transmitted through bacteria or viruses but through prions. Transmission through prions meant that the incubation period could be 5 years, 10 years or possibly even 100 years. This raises the question of how CJD was transmitted in European societies. The amazing answer is that cannibalism was not uncommon in 17th and 18th century England and so it too was probably transmitted through eating people's brains! Mad cow disease was also discovered to be transmitted through feeding cows the ground-up brains of other cows. BSE can also be transmitted through lymph. Lymph is present in blood transfusions and with an incubation period of up to 100 years there could still be horrific consequences in the future from the use of blood transfusions, Maybe Jehovah's Witnesses belief in the evil of blood transfusions will prove to be well=founded.

THE PERSISTENCE OF FAMILY TRADITIONS

I was brought up by German Jewish refugees in an English village. In contradistinction to Daffyd in *L*ittle Britain, who erroneously believed that he was ***"the only gay in the village"***, as far as I know, *we* were **the only Jews in the village** 😀 were extremely laissez-faire and had no interest whatsoever in the fact that at the local primary school which we went to, we sang hymns and carols and learnt to love My Lord Jesus. Religion was rarely mentioned at home. Either my parents had enough "faith" in the intelligence of their children to work things out for themselves or possibly it was that they found it easier just to let their four kids run wild and do what they liked. Another possibility is that they just wanted to blend into the English village life. If we had been practising Jews, there is no doubt that we would have been stigmatised. This does explain why we had never heard of Chanukkah and why we celebrated Christmas. But it doesn't explain why we celebrated Christmas differently from everyone else in the village. Whereas all of our friends eagerly awaited what Santa had brought down the chimney and left in their stockings for Christmas morning, we followed the German Christian tradition of exchanging presents on Christmas Eve.

We were the envy of all our friends as, when darkness fell on Christmas Eve, we would be given our presents, after singing a Christmas carol as per the old 😀

crack of dawn on Christmas morning to ransack their stockings whereas we awoke peacefully and carried on playing with the toys we had received the previous day at 4pm when darkness fell.

To summarise, my parents escaped the persecution of Jews by the Christian society in German and wanted to blend into the English village. For Christmas, they maintained the Christian tradition but the German one which meant that we still stood out as different from all of the rest of the village while continuing the tradition of our Nazi persecutors without even asserting ourselves as Jews.. The only conclusion is that they just carried on doing what they had always done and, in fact, gave no thought to any of my above 😀

I am much more logical and thoughtful than my parents. I am non-religious with children born to my Christian but also non-religious wife and live in

Byron Bay where it is hot and sunny at 4pm on Christmas Eve. Naturally, at 4pm on Christmas Eve we all exchange presents, followed by a slap-up vegan gluten-free feast! Thus is demonstrated the persistence of family traditions.

DEPRESSION WITH NO DISCERNIBLE CAUSE

Patient : Doctor, I'm depressed.
Psychiatrist : Oh dear. Any idea why?
Patient : My friends, family and myself all had small businesses but the government made us all close down although we had done nothing wrong.
Psychiatrist : Anything else?
Patient : We aren't allowed to go to the pub or cafe to discuss our problems.
Psychiatrist : Go on.
Patient : We aren't allowed to visit each other to discuss our problems.
Psychiatrist : I see.
Patient : Actually, I'm not allowed to go anywhere.
Psychiatrist : Hmmm.
Patient : Except vital shopping but then I'm not allowed to go near anybody and, god forbid, definitely not touch or hug them.
Psychiatrist : Oh!
Patient : .. and this could continue into the indefinite future.
Psychiatrist : .. and that's it, well, I'm stumped, I can't fathom any possible cause for your state. And would you please sit a little further away.

ON BEING MILDLY JEWISH

As I said in "The persistence of family traditions", my parents were German Jewish refugees and we were brought up in an English village called Allestree where we were the only Jews in the village. My parents made no friends in the village and had no social life with any of the locals.

My father, Dieter, was a consultant child psychiatrist and he had two friends who lived in nearby Derby: Dr Kirschenberger, who was a Jewish refugee and Dr Conn, an elderly Jewish refugee. Dr Conn was the Director of Kingsway Hospital in Derby, a mental hospital, originally called the Derbyshire Lunatic Asylum, and this is where my father's mother, Olga, ended her days.

Dr Kirschenberger was a very proud G.P. who was caught drink-driving and this refugee, who had escaped the horrors of the Nazis, could not stand the social stigma and he and his wife, Kitty, made a joint suicide pact as a result. This left my father with only one friend, Dr Conn.

My mother, Liss, had many friends. She was very close with her sister and a whole gang of German Jewish refugees but they lived many miles away, in Richmond. They were a happy and sociable bunch but my mother only saw them for a few days a year. By and large these refugees were very close, loved each other, loved their lives, worked and prospered in England and were on good terms with their business associates but, as for actual friendships, these were virtually exclusively between themselves. I never heard any of them express any interest in Judaism. Nor were the Nazis a topic of conversation or a nagging thorn. They were still keen on the German culture of their upbringing.

Later in life, when my parents divorced, my mother went to live in Richmond and from then on had a full and active social life with all of her old childhood German friends especially her sister. My father remarried and moved nearby but, with Drs Conn and Kirschenberger now dead, he never again had a friend.

We four mildly Jewish kids did blend into English village life and not only did we blend but we were something of a hub. We had a large house and garden and laissez-faire parents and this attracted gatherings of friends. This was also the case for my cousins in Richmond. They had the whole top floor of their house to themselves which was completely non-adult territory, except for the weekly charwoman!

Dieter and Liss rarely ventured out into the village of Allestree and never visited or had visits with any locals. We four children rampaged wildly throughout, with many friends whose houses we visited although they were generally rather stuffy and restricting compared to our own. My father was rarely seen as he confined himself to his study which was his sanctum. He was semi-revered when he did appear and talk to our friends. My mother was such a soft and vague and non-judgmental person that she was adored by all our visitors. She wasn't an earth mother who swamped them with refreshments; it's just that they could be their natural selves with no need for English politeness.

My two brothers, my sister and I had no awareness of being Jewish. I still have never known anyone who went to a synagogue or upheld any Jewish customs. The old customs which were upheld were German customs from their childhood but not Jewish ones. For instance, we all exchanged presents on Christmas Eve not Christmas morning and none of my generation had even heard of Chanukkah. We also had treats which were traditionally German but not Jewish, such as pfefferkuchen, apfelstrudel and spekulatius.

I never felt in any way that I was an outsider. I was once called a Nazi by one of my schoolfriends and another used to call me "brown man" but I hardly knew what they were talking about and no-one else took any notice of what they were saying. Although I still feel "English" and more and more mildly Jewish, I am also quite at home living in Byron Bay in Australia. I am now an elderly person and it has only just occurred to me that my parents never did assimilate into English society.

OUT OF SYNC

I started watching **Love Actually** on Netflix last night and noticed straightaway that the vision and sound were out of sync. Then I tried other shows and they were all out of sync. So I spent 65 minutes with Brendan on the Apple technical help line and after resetting everything to factory settings and all sorts of things we discovered that the free-to-air tv was also out of sync, thus proving that it was the tv at fault and not the Apple tv. So Brendan and I reset the whole tv to factory settings and all sorts of other re-settings and everything was still out of sync. So the conclusion was that I could either try LG (the television brand) or replace the tv. Then I tried to watch Youtube on my computer and that was out of sync as well. So I asked Hope to come round and she thought that everything looked fine and in-sync and then I noticed that she was out of sync as well and then Glen came round and he was out of sync as well! I estimated that everything – the tv, the computer, Netflix, Glen, Hope, were all about 200 milliseconds out of sync. Then I got to thinking that light waves reach you about 200 milliseconds before sound waves and so the whole world must be naturally out of sync. I wouldn't be surprised if babies perceive everything out of sync until their brain programs their visual cortex signals with their auditory cortex ones so as to make them appear to be in-sync. For some reason I must have been concentrating too hard and noticing the natural lack of sync and so have reset my brain to factory settings and I expect within a couple of days my brain will have re-adjusted and the world will be back in its illusory state of synchronicity.

BOMDAS AND PEMDAS

More than 50 years ago, when I was at school we were taught how to solve mathematical expressions by the simple rule that you do multiplication and division before addition and subtraction. Yes, you do brackets first but that is quickly taught as being blindingly obvious as otherwise what the hell are the brackets there for. It's also quickly taught that you do an exponent first as an exponent is a part of the number and is also blindingly obvious, not that we knew the word "exponent"; we just used to call it the power of the number i.e. squared, cubed etc. Then in England they brought in the acronym BOMDAS .. brackets, order, multiplication, division, addition, subtraction and the Americans called brackets parentheses and the order of the number the exponent and brought in the acronym PEMDAS. The acronyms were brought in just to make things "SIMPLER" and so that the kiddies couldn't go wrong if they followed the rule of the acronym. But not only is this rule unnecessarily complicated instead of simpler, as I said, once learned it is obvious that you do brackets and exponents first, but to go from the sublime to the ridiculous, the rule is bloody well wrong!! If you take the expression 5-3x7+8 and follow the rule, you get 3x7=21 first, then you do addition and you get 5-29 = -24 and this is wrong!! You should go 5-21=-16+8=-8   In other words, you do everything except the straightforward addition and subtraction first and then work left to right not addition first then subtraction. Yet this is what is being taught throughout English speaking countries and as far as I know throughout the world. You get a lot of people on Facebook trying to solve simple mathematical expressions like this. Most of them have never heard of any of the rules and simply work left to right and get it completely wrong. The majority of people go 5-3=2x7=14+8=22, which is wrong but no more wrong than BOMDAS and PEMDAS. Of those who know BOMDAS or PEMDAS they nearly all get it wrong as well because the rule is wrong. If we come right up to date the kiddies don't use their heads for any calculations, even the simplest ones, like these, enter the whole thing into a scientific calculator and get the right answer without knowing why, because scientific calculators are correctly programmed, not with PEMDAS, as they know that the rule is wrong.

SILLY SALFIELD INTERCHANGES

Nick Salfield (retired director of Derbyshire Health Services)
The Rover was blue and he was nearly killed in a Volvo.
And the Citroen was pink.
Stephen is a rigger head Citroen pie.
Steve Salfield (retired consultant paediatrician)
Ten thousand bolching bolligrobbles, of course.

Philip Salfield (retired psychology researcher)
Breckety brek and an ooh lah lah!

Angie Salfield (retired psychoanalyst) *Oh you are all going prematurely [or not!] senile.........first of all we were all in 1 car...no Simca on that occasion......agree with all the stuff about dad and the Custom's officer......it was very hot and I got sunstroke.....I was about 12 or 13 and you were there Steve. I used to be envious of all the older teenagers getting off with each other. We can ask Clare how old she was when we didn't pick her up! We called it 'biggies' because we were Scottish and piss was 'wee wee' so shit was bigger and we called it 'biggies'- totally rational after all!*

Nick Salfield
Steve if you weren't there when you were 17 or 18, who was that driving the Simca? And he didn't have a prang Phil he drove over the customs officer's foot, had an argument with him and the customs officer did a V sign at Dieter.

Steve Salfield
Oh well Phil is remembering a different holiday than I am, or a conglomeration of various holidays. I was only about 11 or 12 on the holiday I can remember. When I was 17 I was going on holidays

alone with Jacquie I think.

Philip Salfield
Nick was 8, I was 13, Angie was 15 and Steve was 17, and could drive. One day we went out in the car without Dieter or Liss and very excitedly planned a big escape and never to return. We met a French girl called Katerina and the song "Katerina, ah ha ha ha" was our constant refrain. Dieter did nothing but read either outside the caravan or on a deckchair on the beach except one day we were all amazed when he started helping Nick build a sand-castle. Dieter had a small prang with another car when we were queuing up for the ferry and had a furious row with the other driver. We nearly didn't go because Dieter wouldn't get out of bed but eventually with Liss taking him breakfast in bed, doing all the packing and us all pleading with him, we eventually left very late in the morning. Our favourite drink was cacolac. The person I have least recollection of on the holiday is Liss, because I think she spent her whole time trying to please Dieter and to arbitrate between Dieter and us.

Steve Salfield
Well nothing German about biggies, I don't think. It's just what we called it.

Sue Walker

Nick remembers it wasn't a poo he needed...it was biggies! A strange german euphemism perhaps for number 2s, as we english call it!

Nick thinks it wa the holiday you forgot about Clare. Nick said your dad fell through the top bunk in the caravan. He also remembers being on the beach and needing a poo. Dieter took him back to the campsite to use the "English toilets" which you had to pay for because we couldn't stand the smell or mess in the French ones but unfortunately he didn't make it! This is where Nick learnt that it may look further walking on the hard sand, but its always worth it! (having to wipe tears of laughter as Nick related that). He also remembers meeting Chris Tyler and family who 10 years later sold me a non functioning bubble car for £20!

A VISIT FROM STELLA

It's New Year's Eve and Jono and Sara are having a party in their flat in the compound.

I go to bed at midnight and just before I fall asleep, I hear a noise in the house. I switch the light on and there is a little Cairn terrier with a bow in its hair padding towards me and it goes to lie under my bed. I pull it out and check its name tag. It is called Stella, I have no idea whose it is. I ring the number on the tag and then Jono and then Sara but only get answering machines. I had heard all the party-goers leave before midnight to go into town and watch the fireworks. So I take Stella to Jono's house and put her in there.

Jono and Sara return to find a strange little dog lying next to their dog, Oso, on their couch. In the end, they all sleep together in Sara and Jono's bed. In the morning, Sara rings the number on the tag. The owner lives down the road and has lost her dog, which escaped when she was out at a party.

Sorry, but that's the end of the story. The missing links to the story will forever remain a mystery.

COVID19 APPRAISED STATISTICALLY

Global population   7700 million
Annual global births   130 million
Annual global deaths   56 million
Annual global road injuries   50 million
Annual global road deaths   1.25 million
Global cases of COVID-19   1.12 million
Global deaths from COVID-19   0.064 million
Australian cases of COVID-19   0.005 million
Australian COVID-19 deaths   0.00034 million

May 2nd 2020:
Global COVID deaths over the last 5 months = 150,000
Global deaths over the last 5 months : 20,330,000
Percentage of world's population dying from COVID=0.002%
OR percentage not dying form it = 99.998%
OR out of 100,000 people 2 have died from Covid in 5 months
OR for every person who has died of COVID, 120 people died of other causes.

May 22nd 2020
There have been 5 million cases of COVID-19

worldwide.
330,000 people in the world have died of COVID-19 to date.
7,699,670,000 have not.
28,000,000 died of other causes.

The world population is 7,440,000,000 which is 7.4 billion. If 80,000,000 (80 million) people died of COVID-19 the world population would be 7,360,000,000 and we would then say the world population is still 7.4 billion. In other words, 80 million deaths from COVID are completely insignificant statistically.

This is what you can glean from the statistics that I have presented.
First of all, it puts the danger of COVID-19 into perspective.
Secondly, the majority of human beings simply do not understand numbers.
Thirdly, as Mark Twain said, there are *lies, damned lies and statistics*.
Fourthly, the danger from a pandemic shrinks into insignificance compared to the danger from environmental degradation and yet the effort and resources are diametrically opposite to this logic.

THE STANDARD MODEL OF PHYSICS

In the standard model of physics everything in the universe is made up of just two fundamental particles, quarks and leptons. The generic term for them is hadrons. Hence the Large Hadron Collider LHC, which is the most significant part of the CERN nuclear research facility in Switzerland. A hadron is a subatomic particle such as a baryon or meson. A baryon includes nucleons and hyperons. A meson is intermediate in mass between a proton and neutron and transmits the strong force. The strong force is carried by gluons which act on quarks (which include electrons) to give them mass, such as photons and neutrons. The weak force is carried by the Higgs boson, which may not exist but if it does it annihilates immediately when it is produced. The Higgs boson acts on leptons to give them mass. Unfortunately, none of this explains gravity which so far remains outside the standard model and can only be understood by Einstein's theory of general relativity.

I thought it would be a good idea to give a quick simple summary so that we are all on the same page in

THINGS I DON'T BELIEVE IN

The Big Bang Theory
Creationism
Darwinian evolution
All religious doctrines
Pantheism
The Soul
Past Lives
Reincarnation
God
Science

UNSOLVED MYSTERIES OF THE UNIVERSE

Pre Big Bang Event Horizon

The Origin of the Universe

The Origin of Life

The mechanism of evolution

Ball Lightning

Dark matter

Dark energy

Life

Death

Wormholes

Cure for the common cold

Consciousness

Time

The dimensions of a reference point

What is a straight line

What is up and down

WHITE COUPLE MAKING LOVE

I was sitting on my verandah and not ten metres away there was a white couple pecking and canoodling and rubbing each other up for about half an hour. Then he mounted her and thrust in and out for about two minutes. Then he got off. They gave each other a few more pecks and cuddles and then they flew off.

THE CASE FROM HELL

In late November 2011 Jack Stokvis of Haworth, New Jersey, received an unusual email stating that he was a beneficiary to an unclaimed inheritance in Australia from his paternal great grandfather, John Meyerfeld (1838-1907) who died 104 years earlier. The email came from a reputable international heir search company in Germany and stated that they got Jack's email address from Horst Schröder with whom Jack had corresponded years before. They claimed that John Meyerfeld's inheritance came from a recently sold parcel of land in New South Wales. Jack was naturally suspicious that this could be another of the many scam emails he receives but was also curious as the letter and follow-up telephone calls gave accurate information about his grandmother Alice Meyerfeld Stokvis, her sister, Edith Meyerfeld Barry, their brother Curt Meyerfeld, and Jack's parents. Jack's brother also received the suspicious email and thought that it was probably "BS."

Jack was determined to learn the truth. He emailed his second cousin, Oliver Chamberlain, asking him to ask his 91 year-old mother, Nell, if she knew anything about it, Nell wrote a letter to Horst Schröder, with whom she had been corresponding about another family member, Max Meyerfeld, who had been a translator for Oscar Wilde's works. Meanwhile, Jack got in touch with the rest of his extended family in the UK, the USA and Australia about this unusual story. This is John Meyerfeld.

Jack is my second cousin and as I live in Australia, I volunteered to become the lead detective and enlisted the services of my high school friend, Rosemary Annable, who was working as an archivist in Sydney. Jack nick-named her "the sleuth" because she confirmed that there was indeed an unclaimed inheritance in the name of a John Meyerfeld who died in 1907. The money was being held in a trust fund set up by a firm of Sydney solicitors. It turned out that one of the solicitors was a descendant of Hon. George Henry Cox who along with Samuel Aaron Joseph were John Meyerfeld's partners. He was also a beneficiary from the property sale. The property sold for A$800,000 and John Meyerfeld owned a quarter share.

We all hoped that it would not take long to receive the money because we wanted 91 year-old Nell to receive her share of her grandfather's estate. Unfortunately, it took eight years for the bumbling legal system to finalise the case before the money was disbursed in 2019 and sadly Nell had died in 2013 and never received anything.

The five lots had been sold in 2008. Chris Connellan, who called it *'The case from hell'* when we phoned him told us that his partner, Mr Cox had received his share of the proceeds and that John Meyerfeld's share of the sale proceeds had been secured in a trust fund, awaiting a claim from the rightful inheritors!!

In 1886, after spending fifteen years in Australia, John wrapped up Meyerfeld & Co and returned to Germany, a wealthy man, at age 48. He married Jenny Lippmann in the same year. Jenny was 23 years old from Aachen.

This is the schooner, Loch Lee, one of John Meyerfeld's three trading ships.

John Meyerfeld died on 13 September 1907, aged 69. He lived at fashionable Kurfürstendamm in Berlin. His will left everything to his wife Jenny and his two surviving daughters.

Probate for the John Meyerfeld estate was granted on 5 December 1912 at the London Registry. His effects were valued at £ 486. 12s. His will was lodged in an English translation at the Leeds Registry.

We assume that he had forgotten all about the acquisition of the trivial parcel of land near Lake Macquarie and hence the unclaimed inheritance when it was sold in 2008, his quarter share amounting to approximately $200,000. Which brings us to the year 2011. The year 2011 was drawing to an end when Jack Stokvis received an email stating that it had been recently discovered that he was a beneficiary to an unclaimed inheritance dating back to 1907.

DEATH IN A TIME OF CORONA

"Nothing in this life is so certain as death and taxes". Obviously the taxes part is meant as a joke and not necessarily true, but the death part is particularly relevant to the COVID-19 crisis and is constantly misunderstood. This is the essence of the problem we have in this time of COVID-19 : people in our society are scared shitless of dying. This is the most unfortunate state of affairs that could possibly exist. We are all going to die and so to have this terrible fear hanging over you for the whole of your life ruins every aspect of your life.

It is sad when people die before their time BUT fellow peers of my generation and beyond – we who are now 70 years old or more – we should welcome the advent of death. We have lived our lives, we have had our trials and tribulations, we are living through the travails of the present and before us is the wonderful release of death, where nothing further can trouble us.

There is nothing to fear. It is pure oblivion. Purer, even, much purer than the oblivion you experience in the deepest of sleeps or meditations. Why would you seek to carry on for as long as possible into the world of helpless old age, the only respite being your next medical appointment.

We should make the imminent advent of death a thing to rejoice in, to allow us to completely relax and have nothing to worry about as we enjoy each of our remaining days.

Let the younger ones worry about their health, their finances, the destruction of the Great Barrier Reef and the rainforests, the spreading of viruses, the over-populated world. Perhaps the blame lies with our generation .. too bad, we are about to die and that will be the end of it for us.

This philosophy applies directly to the current ridiculous COVID -19 crisis. All efforts should be focussed on the younger generations, on making sure that they have a continued livelihood and liveable environment, and medical treatment in the rare cases where that should be needed. Forget about us of the older generations. We have had our time. Leave us be to get it or not, to pull through it or not, but don't deprive any of the younger ones for our sake.

DO I BELIEVE IN GOD?

Every now and then someone will ask me if I believe in God or if I am an atheist or an agnostic. At first glance it sounds irrefutably logical that I must be able to choose one of these answers : yes, I believe / no, I don't believe / I don't know if I believe or not.

That sounds pretty easy, you're thinking, if you don't know if you believe in god or not you're obviously an agnostic.

But this is not so.

It's not that I don't know what I believe. What I believe is that there is some fundamental mystery that I will never comprehend. Some people might call this mystery "GOD" but I don't think it is actually what most people would call God or if that is what they would understand by my answer if I said that I did believe in God.

It depends on how you define God.

I could categorically tell you I am an atheist if you're talking about any of the gods I have ever heard mentioned, all of which exist within our ideas of time and space. But then if you define God as "some totally other reality" then you might say I am a believer, but I think this would give most people a false idea of what

I believe because what I am saying is that I can see no way to ever understand the fundamental mystery of the universe not that I believe yes here is the universe and there is God.

The question I want to answer is whether or not I believe there is some fundamental and unfathomable mystery to the universe and I believe that for me as an ordinary limited human being that there is. Even if there should turn out to be this weird thing called God in some bodily or spiritual guise wafting around the universe creating and destroying things it wouldn't answer the mystery to me of what was there before the beginning of the universe and where does the universe end.

If you ask me if I believe in God you are making all sorts of assumptions. You seem to be assuming that our universe is a real place and asking whether there is anything beyond this in some way that we can call God. And I think that is where I am an agnostic – I don't know if the universe is real or not but I veer towards thinking it's not.

If I believed that the universe is a real phenomenon then I would be a pantheist because I had a look at the website of the World Pantheist Movement and it is a very intelligent belief system with many members who are much cleverer than I am, unlike adherents from other religions whose beliefs I find ridiculous. I would agree with the pantheists that everything in our universe is connected, that there is no separate being called God and that there are no messiahs and I would also agree that everything can eventually be explained by science.

But I do not believe that our universe is real. Otherwise how can it both be impossible for it to begin and impossible for it not to begin. And impossible for it to end in space and impossible for it not to end in space. So, unfortunately I cannot subscribe to being a pantheist.

Something is wrong.

TIME AND SPACE CANNOT BE REAL

The only way I can understand anything is in terms of time and space.

Maybe the answer is that the only real thing is ME and absolutely everything is my imagination. But this doesn't solve anything as I still exist in time and space. So the mystery would still be there.

Quantum mechanics, quantum physics and astronomy are very amazing at increasing knowledge and have amazingly clever theories of the beginning of the universe and the beginning of time and when the universe will end and how life started and all sorts of things.

But all of this is still based on the concepts of time and space. There has to be a beginning to the universe and there has to be an end – both in time and space – because it is impossible to have something without a beginning and an end whether it is God or the universe or anything – but it is also impossible that the universe has a beginning or an end. So logically I don't think that the universe is real. In fact, I wouldn't be surprised if quantum physicists themselves come to this conclusion.

It's not really that important to me whether God exists or not because even if there is a god, that God still exists in time and space because I cannot conceptualise anything else and it would not explain the mystery of how the truth is beyond time and space. Even if God is energy or a spirit that permeates everything there it is in time and space in the universe.

So I claim to know the answer to the question of God and the universe and that answer is that there is a

MYSTERY

something

INCOMPREHENSIBLE

something

INEXPLICABLE

and not just now until physicists or some spiritual master discovers the secret of the universe, but for always, always unavailable because time and space must be real and yet time and space cannot be real – even as I say "always" that itself is a concept of time which in itself indicates that it is the correct conclusion that there is an incomprehensible inexplicable mystery.

because TIME AND SPACE CANNOT BE REAL

and yet WE LIVE AND THINK ENTIRELY WITHIN TIME AND SPACE

In conclusion,

next time I am asked whether I am an atheist, an agnostic or a believer, I will say

I AM ………..

an atheist who believes there is a fundamental mystery

and

a believer in God but not in the normal sense

and

an agnostic in that I don't claim to know the ultimate answer.

MY FINAL COMMENTS ON THIS FUCKING CORONAVIRUS

I would prefer to get COVID-19 and die of it and for 20% of the world's population to die as well rather than live in a totalitarian state. But, of course, we are not now living in just a totalitarian state we are now living in a totalitarian world. The world leaders are jubilant. I am referring of course to Donald Trump, Vladimir Putin, Xi Jinping and most other countries' leaders as well. They are bitter, narcissistic megalomaniacs, hoping to see out their chauvinistic old age in a glory of power.

What a brilliant coup! I have to give them credit for their brilliance. They have the entire population of the world cowering in dread fear, socially isolating i.e. looking after number one and no-one daring to come out and speak against their policies.

THE BRAIN'S WAY OF HEALING

In "***The Brain's Way of Healing***", Norman Doidge brings us up to date with the advances in neuroplasticity. The paradigm of modern Western medicine is of the doctor curing the patient with specific pharmaceuticals. In the paradigm of neuroplasticity, it is the doctor's job to train the patient to cure him or herself with a lot of hard work over long periods of time, not to be a passive receiver of pharmaceuticals from an omniscient doctor. "Cure" is in fact the wrong word as neuroplastic techniques will generally relieve the symptoms but not eradicate the disease. His favourite example is a man with Parkinsons disease who has remained virtually symptom-free over 40 years, although he still has Parkinsons. In neuro-degenerative diseases like this, it is not the muscles which are malfunctioning, but the neuronal signals. Walking and the other movements which degenerate in Parkinsons are controlled by the autonomic nervous system. But it is possible to learn to control the movements with the higher areas of the brain such as the cortex, to ingrain the movements in these areas of the brain with huge amounts of persistence and practice and thus to show no symptoms. As for chronic pain, it is not located in the body part but in the brain. The brain has become accustomed to these pain circuits and so the chronic pain continues, which unlike acute pain, is thoroughly debilitating and of no value. So the brain has to be trained to discontinue these habituated pain circuits. It all depends on which part of the brain is responsible for the feelings of pain. If the pain circuits are in the visual cortex, then the circuits can be disconnected by replacing them with visual experiences. So when the patient experiences pain, he or she must concentrate on visualisations. One patient who had had continuous chronic pain for 20 years, spent 8 hours per day visualising for 2 months and became permanently pain-free. The alternative treatment is to use strong pain killers which have debilitating side effects whilst their effectiveness continually deteriorates, thus requiring stronger and stronger doses. The neuroplastic healing technique has similarities with meditation and self-hypnosis but it is distinctly different. It requires specific concentrated thinking as opposed to ridding the mind of thought. However, all three practices are holistic in the manner of traditional Eastern medicine as opposed to the reductionist paradigm of Western medicine. The way forward needed by Western doctors should be to help patients heal themselves with longer appointments with detailed instruction requiring long hours of dedicated work from the patient and not with a deceptively easy fix of pharmaceuticals. It is unfortunate that the dominant influence of pharmaceutical companies, their relationship with doctors and the constraints on a doctor's time, all make it unlikely for this to happen.

REDUCTIONISM HOLISM AND MU

Whether you are thinking about medicine, maths, art, music or philosophy, there are three levels at which these and other subjects can be considered. Douglas Hofstadter devised the following diagram, which I found to be very helpful in understanding this concept. If you look closely, you will find the three words : **HOLISM,** spelt in the alternative way without the **W, REDUCTIONISM** and **MU.**

First is the *reductionist* level such as a doctor diagnosing someone with high blood pressure and prescribing a drug which reduces it. The next level is the *wholistic* one, usually spelt "holistic" but I prefer the former spelling as that is exactly what it means. In this case, the healing practitioner would consider the person's whole constitution and lifestyle and prescribe a whole raft of remedies from diet to exercise and work-life balance. The *mu* level incorporates both the previous two levels and would recommend all the above approaches.

The three levels can be seen in this example from the art of Escher.

In his book about reductionism, wholism and mu, Douglas Hofstadter has chosen the greatest exponents of the concept in the three fields of philosophy, art and music. These are Gödel, Escher and Bach, Hence the the title of his book : **Gödel, Escher, Bach.** The message is that things exist on all sorts of different levels at the same time which is the mu and that the sum of the reductionistic levels does not equal the (w)holistic level and vice versa, but both levels are true at the same time, it is just a matter of where your consciousness is focussed.

CONSCIOUSNESS BEYOND LIFE

This is the title of a book by Pim van Lommel, who is a surgeon who has operated on people who were completely unconscious, who showed no neural activity, their eyes and ears were taped and yet when revived after the operation they reported that they were on the ceiling looking down on the operation and they have correctly reported procedures of the operation and conversations that were going on. Pim van Lommel believes that aspects of consciousness correspond with the basic principles of quantum mechanics such as non-locality, entanglement and inter-connectedness. Communication can occur instantaneously in a timeless and placeless dimension. Sub-atomic particles can affect each other instantaneously at any location in the universe. He believes that our physical realm is just one manifestation of non-local consciousness, subject to the processes of quantum mechanics whereby sub-atomic particles are sometimes waves and move in probabilities depending on the observer – they operate beyond the realms of time and space.

DO WE REALLY EXIST?

If we condense Descartes' philosophy to cogito ergo sum (I think, therefore I am), this tells us very superficially and briefly what Descartes thought was proof that we exist. However, in his book, **Descartes' Error,** Antonio Damasio puts forward the thesis that cognition is not merely a logical process but also an emotional one. He provides detailed neuro-physiological descriptions of how neural circuits in many different parts of the brain, the central nervous system and the peripheral nervous system, combine with hormonal and other chemical secretions from glands and parts of the brain which are then transmitted via the endocrine system as well as the immune system, all forming continuous feedback circuits and somehow ending up as thoughts, ideas and action; the source or origin of all this activity being one's somato-sensory system either externally, internally or both. The point he is making is that the proof of one's existence is not as Descartes said just through the logic of ones' senses but that emotion as well as cognition is equally important and that this comes from one's interaction with the environment, proving that the world and hence oneself do definitely exist. I have another bigger error to do with Descartes which is that his ideas all stem from the primacy of the acceptance of the existence of the Christian God.

However, even though I think that it is ludicrous for Descartes to take the existence of the Christian God as implicit, I think it is a huge omission on Damasio's part to ignore the spiritual aspect of the human being. Neither Descartes nor Damasio deal with the final holistic level of the human being which is that above and beyond the mind there is the self, which is spiritual rather than somatic in nature and that this is the connecting point with what I have termed as universal energy.

THE UNIVERSE IN YOUR HAND

This is the title of **Christophe Galfard's** book about the world of theoretical physics at the beginning of the 21st century. We learn that there are electrons in the outer space vacuum that are moving closer to each other with virtual photons popping out of the electromagnetic field and then both the electrons and virtual photons are gone!! We then discover that the visible universe only contains 4.6% of the energy of the universe, the rest being dark energy and dark matter, which are still complete mysteries! This was the first time in which I, personally, read of the theory that the Big Bang Theory, about which I have always been disparaging and which I spent a long time dismissing in my own book about the universe, which we all know from Stephen Hawking was the beginning of the universe, is no longer considered to be the beginning of the universe by the theoretical physicists themselves. In fact, my disparaging remarks become even more pertinent when we discover that the Big Bang is not the beginning of the universe but rather the beginning of *spacetime*. Spacetime was preceded by the *inflationary epoch*, separated from the Big Bang by the Planck Wall, which lies 380,000 years before the surface of last scattering!! Yes, that's right 380,000 years prior to the beginning of time!! In my book ***"God, Humanity and the the Mystery of the the Universe"***, I was disparaging about the Big Bang Theory because it was oxymoronic, in that time is how we judge when something starts and the very title of Stephen Hawking's book, *"A brief history of time"*, demonstrates a fundamental contradiction. I must admit that I was unaware of the Planck Wall and that prior to the Planck Wall and the last scattering, all was opaque and there was no light, no time and no space. Now, that has got to be the biggest oxymoron that the physicists have ever come up with, a period of time before time even began!! All of which, makes me more confident in the conclusions in my book, that these ideas are oxymoronic and that the full truth will never be discovered through the world of science. One thing is certain and that is that these theoretical physicists such as Christophe Galfard, Stephen Hawking, Albert Einstein, Richard Feynman and Max Planck have the most brilliant and imaginative minds. For instance, the current theory in vogue is string theory. In this, the constituents of the universe are entirely made up of both straight and closed strings in 10 dimensions and this means that there are either many universes or maybe even an infinite number. I certainly do not have the hubris to even begin to compare my intelligence with theirs. However, I do believe that their brilliance is constrained by their implicit belief in the supremacy of science and particularly physics, whereas I believe that I can see the wood from the trees more clearly than they can and whereas they still think that a Grand Unifying Theory can be found in the conscious state through the world of science that my conclusion makes more sense that in the last analysis all is ineffable. The one physicist who is possibly an exception to this is Albert Einstein, who maybe realised that science is not where the final truth lies.

DOROTHY'S FUNERAL

This is Dorothy's funeral. Watching on are her chimpanzee friends and family. "Her facial expression was serene, one of final repose". This is from the book **Kindred Beings** by **Sheri Speede**. Raymond conducted a funeral service for the staff, volunteers, and dozens of people from the village community. He spoke of Dorothy's suffering and of the happiness she had known at Sanga-Yong Chimpanzee Rescue Center. He told the community that the best way to honour Dorothy would be to never eat the meat of chimpanzees again   Soon, chimpanzees will be extinct, for a variety of reasons but principally because humans continue to eat them.

BIOCENTRISM

Biocentrism is **Robert Lanza's** theory of reality which is expounded in his two books with **Bob Berman**.

The most revered name in quantum mechanics is **Richard Feynman** and he is quoted as saying "If you think you understand quantum theory, you don't understand quantum theory." I would add that the same applies to biocentrism.

The weird world of quantum phenomena is now an accepted part of modern physics. Sub-atomic particles only come into existence when they are observed. Prior to this they are waves of probability. This has been demonstrated in the famous double slit experiment. Other weird aspects of sub-atomic matter are superpositionality, which means that it doesn't inhabit any specific place – it only does this when it is observed and its probability wave collapses. When we say it doesn't inhabit any specific place it is not known whether this means it doesn't inhabit any place or it inhabits every place in the universe. We are already in the spooky world of quantum phenomena but, on top of all this, the wave function which becomes a sub-atomic particle does not inhabit any particular time either.

In quantum theory, as postulated by **Max Planck** and **Niels Bohr,** electrons move in orbits round a nucleus and when they move, they move directly to another specific orbit, which is called a quantum; if this is an inner orbit then they emit a photon, and that is what light is. When they move from the outer orbit to the inner orbit they do not pass through the intervening space, they just reappear in the new orbit. Lanza takes this logic even further. He says that an electron does not orbit a nucleus, rather, it shimmers at a likely distance in a state of probability until it is observed, which forces its wave function to collapse and it then adopts a particular orbit and position.

And still getting weirder. If a light is shone into certain crystals two identical photons can be formed, which fly away in different directions, but they secretly share a wave function. If one is observed, its wave function collapses and it becomes an entity. With no time delay, and even if it is at half a universe's distance away, the wave function of the twin then also collapses but it assumes the complementary guise of its twin; if its twin had a vertical spin it adopts a horizontal spin and vice versa. This is called quantum entanglement.

And yet weirder. If sub-atomic particles are observed, this can also alter things which had happened previously.

These quantum phenomena can only mean that neither time nor space are real external phenomena. Eyes receive light waves, ears receive sound waves, touch receptors in the skin receive various other forms of input. But, there are no receptors for time or space. Time and space are human constructs. Time makes sense of the process in which things change and space allows objects to be separate and distinct. Which brings us to how we perceive the external and, for that matter, internal world. Our sense organs receive waves of probability but it is only once this information is transmitted to the brain that a perception occurs and this is when the probability wave collapses and becomes a perceived reality. The reality of anything in the universe depends on an interaction between a probability wave and an observer. It is reasonable to suppose that the same spooky world of sub-atomic phenomena also applies at every level of reality. After all, every macro object and organism consists of sub-atomic ingredients. At first, this seems completely unbelievable as we have amazing scientific and technological developments which are real and work and we have historical records which show past events as they actually happened and all agreed upon by all the human observers. But all of this is due to the fact that the probability waves have been observed by conscious entities and so the waves have collapsed into physical realities. Without the observers, there would be nothing. The reason why there is no apparent weirdness in the macro universe is that there is no way of observing phenomena before they are observed! The theory is that none of it exists without being observed. How could the universe or anything be conceptualised without an observer. However, given human ingenuity, in the same way as the double slit experiment has convinced scientists that this quirky phenomenon holds true at the sub-atomic level, maybe some way will be discovered to demonstrate this at the macro level.

One may think that we already have a good model that makes sense of the universe, so what is this rubbish about turning everything upside down with cranky new weird theories which make no intuitive sense to anyone in their right mind. Surely, the universe began with the Big Bang, the earth was formed, life arose and evolved according to Darwinian theory and consciousness was achieved. However, physicists have found it necessary to introduce the **Inflationary Epoch** to precede **the Big Bang** which is separated from the Big Bang by the

Planck Wall – this is not a joke! The Darwinian theorists have found it necessary to move on to ***Neo-Darwinism.*** No-one has come up with a meaningful explanation of consciousness. Meanwhile, the universe, which is supposedly everything, continues expanding into the nothingness surrounding it! So, as it makes perfect sense that time began at no time and matter came from nothing, we clearly need no alternative theory!! Furthermore, the mathematical physics of the universe all adds up perfectly except for a small question of 96% which is postulated to be dark energy no-one expects to be ever detected! And just a couple of other small problems – no-one has come up with any plausible explanation of how life originated or how Darwinian evolution actually works.

The prevailing scientific doctrine of today is holding on to these tenets in just the same way as religious adherents hold on to theirs, despite all the evidence to the contrary and despite the evidence to the contrary having been discovered within their own discipline of physics.

The theory of ***biocentrism*** provides an alternative to the universe arising with the Big Bang from nothingness and life appearing on earth due to the Goldilocks phenomenon. And we really do desperately need an alternative theory to these nonsensical ones. It was **Fred Hoyle** who coined the term the *"Big Bang"* in 1949 which he intended to be a pejorative term to describe how ridiculous an idea it was. Unfortunately, he was the only one with a sense of humour! The theory of biocentrism that reality depends on the interaction of a conscious observer and waves of probability does also sound nonsensical. However, it is in keeping with the known findings of quantum mechanics. The conscious observer is primary, I won't say first because that involves the concept of time and only then does the universe and everything therein collapse into reality. Without the observer there is no universe and no reality, just waves of probability.

There is a coherent world of science and technology which works within these parameters of time and space, but it cannot make sense of the fundamental questions. Physicists and biologists should acknowledge that the current theories and explanations are useful in many ways but that none of their axioms can be the absolute truth because they are all based on the concepts of space and time. These are artificial human constructs and do not hold up for all the most fundamental questions .. the origins of the universe and life and the process of evolution. The prevailing theories of modern science should stick to their role in providing new technological and medical advances – the probability waves have been resolved into reality and make sense at that level.

The theory of biocentrism solves all of these otherwise insuperable problems and has huge implications for the ideas of life, death and consciousness. However, in my opinion, it by no means resolves our understanding of everything and it seems very unlikely that humans can be capable of fully understanding the universe. Nonetheless, Robert Lanza is absolutely correct that biocentrism, which holds that consciousness is central and that the universe follows on from this, fits the known scientific discoveries much much better than the currently prevailing theories of the Big Bang and the origins of life and the process of evolution. The main advantage of biocentrism is that it does not depend on the concepts of time and space.

Lanza and Berman list the following principles of biocentrism :

. What we perceive as reality is a process which involves our consciousness. An "external" reality, if it existed, would – by definition – have to exist in space. But this is meaningless, because space and time are not absolute realities but rather tools of the human and animal mind.

. Our external and internal perceptions are inextricably intertwined. They are different sides of the same coin and cannot be divorced from one another.

. The behaviour of sub-atomic particles – indeed all particles and objects – is inextricably linked to the presence of an observer. Without the existence of a conscious observer, they are at best in an undetermined state of probability waves.

. Without consciousness, "matter" dwells in an undetermined state of probability. Any universe that could have preceded consciousness only existed in a probability state.

. The structure of the universe is explainable only through biocentrism. The universe is fine-tuned for life, which makes perfect sense as life creates the universe, not the other way around. The "universe" is simply the complete spatio-temporal logic of the self.

. Time does not have a real existence outside of animal-sense perception. It is the process by which we perceive changes in the universe.

. Space, like time, is not an object or a thing. Space is another form of our animal understanding and does not have an independent reality. We carry space and time around with us like turtles with shells. Thus, there is no absolute self-existing matrix in which physical events occur independent of life.

When comparing Darwin's theory of random mutation to explain evolution the counter-argument is how long would it take a million monkeys to type the works of literature : it would take 36 trillion years to write the opening line of Moby Dick : "Call me Ishmael". How many ways are there of arranging 10 books on a shelf? : 3.6 million : the chances of putting 10 books on a shelf in alphabetical order by chance are 1 in 3.6 million.

If reality is created from a human mind, then the human brain must be some piece of work and indeed it is. In a similar vein to Feynman's paradox about quantum theory, George E Pugh has made a paradox about the human brain : "If the human brain were so simple that we could understand it, we would be so simple that we couldn't". Contrary to the commonly held assumption, the brain is not a digital machine but an analogue one – the connection between neurones is not an on/off switch, it is a connection which varies in intensity and frequency – also, the neurone is receiving stimulus from several others, some of which cause excitation and some suppression – so it is a complex symphony of modulated strengths. The nervous system is even more complicated than this but these facts are sufficient to belie the idea of a digital system.

What are "solid" objects composed of? - electrical fields – and what is the feeling of a solid object? - it is the repulsion of your own electrical field to that of the object – nothing is solid, it is just all energy fields and the energy fields exist only in the detector .. your mind. "Indeed, the universe can be viewed as a blurry, probabilistic state of potential information, which the mind-system "collapses" into actual information and sensations" - this is what bestows the feelings of a "me" - so is this what consciousness is? ..

In conclusion, the physicists of the origins and nature of the universe do not appear to be communicating with the quantum physicists. They do not appear to be taking on board the ideas of quantum theory. String theory is the attempt to make a GUT (Grand Unified Theory of everything) based on the mathematical fiddling of one-dimensional strings many trillions of times smaller than sub-atomic particles and employing about 8 extra dimensions to make the maths work – all completely untestable and with no basis in reality. Biocentrism, although still not solving the fundamental mysteries, at least makes coherent sense within the latest findings of physics.

Interestingly, the theories which have most similarities with biocentrism are to be found within the tenets of Hinduism and Buddhism. Biocentrism and Buddhism are superior to the current Western theories in that they do not rely on the constructs of time and space. Unfortunately, as with the Big Bang relying on an inexplicable source which physicists call a singularity, and with religions relying on an inexplicable phenomenon they call God, so too does biocentrism rely on the inexplicable origins of consciousness as well as the waves of probability. So, despite all of this brilliant conjecture, it is my opinion that we still remain entrenched in utter mystery.

Major historical events of the modern era:

- Industrial Revolution
- The Great War
- Second World War
- Cold War
- Imminent extinction of mankind through climate change
- The Great Toilet Paper CRISIS OF 2020

A PICTURE TELLS A THOUSAND WORDS

Donald Trump

The Archbishop of Canterbury

Pope Francis

Mother Teresa

The Dalai Lama

Mahatma Gandhi

THE AUTHOR

I was born in the UK in 1950. My early upbringing was next door to a farm where they reared animals for slaughter, all set in the grounds of a mental hospital. I have studied medicine, psychology, education, nutrition and educational toys. I moved to Byron Bay in 1986. I started writing books about twelve years ago, mostly on the general theme I guess of philosophy.

www.ingramcontent.com/pod-product-compliance
Lightning Source LLC
Chambersburg PA
CBHW040517220526
45473CB00012B/2892